The Terraforming and Colonisation of Venus

Charles Joynson

ISBN:
ISBN-13: 978-0-9956741-5-8

The love for all living creatures is the most noble attribute of man.
Charles Darwin

1 - INTRODUCTION

For hundreds of years people had looked at Venus and wondered if it could be made into a home for humans. However Venus had a tremendously hostile environment; it had a superheated thick carbon dioxide atmosphere with enormous surface pressures and wind speeds of over 300 kilometres per hour. The only way to even begin to make Venus habitable was to remove or dramatically alter its atmosphere, but that was impossible.

But still people continued to cast their greedy eyes at Venus's valuable real estate. Mars had been an easier challenge to make habitable for human life; it had only the thinnest of carbon dioxide atmospheres which needed supplementing rather than replacing. Mars' tilt and day length were very close to Earth's making Mars a much easier planet to terraform. Venus was very different; it had a day length of over 100 Earth days, it rotated clockwise unlike most of the other planets in the solar system, it had a 3 degree tilt which meant it possessed no seasons, and it was highly volcanic.

Conferences took place in 2623 and 2633 to look for ways in which Venus might be tamed. However, because of the hostile nature of the planet, realistic options which would not damage the rest of the solar system were hard to find.

A number of plans were proposed. The first suggestion was to seed the upper parts of the atmosphere with floating plants which would excrete oxygen into an internal bladder. These could then screen the planet from the heat of the sun and as they multiplied build a layer which would grow to cool and consume the whole atmosphere.

A second strategy was to bombard the planet with minerals such as calcium in the form of asteroids which would lock away carbon as calcium carbonate or some other combination.

The final idea was to create a verse-hole or worm-hole to remove Venus's atmosphere. Without the removal of this superheated and toxic shroud around the planet, no life would ever survive on its surface. The proposers believed that they could create such a thing, that it would remove the atmosphere permanently, and that there would be no risk of it polluting the rest of the solar system with waste gasses.

Other ideas were mooted, but were dismissed for being unrealistic or for bearing the risk of damaging the rest of the solar system. Many of the ideas voiced involved plans to strip Venus' atmosphere and let it loose amongst the planets. This carried the risk that these gasses might then be captured by Earth or Mars, thereby warming or toxifying their atmospheres.

Of the three more realistic ideas, there was still plenty of criticism. The Venus floating plant proposal was considered to be far too slow as it might take tens of thousands of years to work. A second criticism of this idea was that only the top few metres of the plant layer would be exposed to the sun and any that fell below this layer would be incinerated by the planet's superheated atmosphere. A third criticism was that the plants would need water which was almost absent from the Venusian atmosphere.

In the case of the calcium asteroid concept there was a lot of doubt as to whether there was enough calcium in the solar system to do the job. It was known that there was some calcium in the Asteroid and Kuiper belts, and it was suspected that there might be some in the Oort and Hills clouds, but calcium asteroids in these places were believed to be too far apart and rare to be a usable resource for an atmospheric sequestration programme. Calcium did exist on Venus but was already locked away in volcanic minerals, so the addition of any calcium would help both living things and atmosphere. But there simply wasn't enough calcium available from other sources to completely transform Venus' atmosphere. Researchers did suggest the use of other minerals which would do the same job, such as fragmented iron; yet none of the suggested substances were present in sufficient quantity or were too distant to be easily obtainable.

The criticism of the verse-hole idea was that it was completely impossible and even if it did work, it might swallow the whole planet rather than just its atmosphere. In spite of this criticism, the scientists who had suggested the idea proposed building a demonstration of a one way verse-hole in the Kuiper belt to prove the theory. They said they would use it to remove a small moon from its orbit around a dwarf planet, without affecting the larger body. In achieving this they could prove their ability to control the verse-hole and avoid any risk to the solar system.

Further conferences took place focusing on Venus's long day and its minimal rotational tilt. All of the proposals made involved using asteroid impacts to modify these. The maths and feasibility of these strategies were far more conventional than the atmosphere extraction plans. They did, however, require the atmosphere to be removed or captured before any asteroids hit Venus.

The verse-hole experiment took place in the Kuiper Belt in 2640. The moon was just two-hundred metres across and the dwarf planet twenty-five kilometres, so being able to remove the moon without losing the dwarf planet was identified to be a good test of control. Intersecting particle beams were focused on a cube of catalyst material which had been placed on the surface of the moon. Once the beams had been triggered, the moon was seen to disappear as predicted. The hole was closed successfully with no signs of any negative effects, and the dwarf planet continued in its slow orbit around the sun. The experiment was therefore deemed a complete success.

After the Kuiper Belt experiment, the calculations for the Venusian verse-hole project were repeatedly re-calculated and re-evaluated. Each step in the extraction process was debated and modelled again and again until the vast majority of scientists believed it would succeed.

Atmospheric extraction was planned for January 2651 when Venus was on the opposite side of the solar system and as far away as possible from Earth and Mars. Sixteen ships were to control the particle beams to ensure that at least three would have contact with the catalyst at any one instant. The beams were synchronised to intersect in the upper atmosphere and to track toward the surface over the course of twelve hours as the planet Venus moved.

At zero hour, the catalyst was dropped into the high cloud layer and seconds later the beams were activated. Onlookers saw a furious typhoon appear in Venus's atmosphere. As the verse-hole did its work, they saw the swirling cloud expand to wrap itself around the entire planet. The white clouds were gradually recoloured red, brown and then black as the extraction point approached the surface. All the while the particle beams had to be continually refocused to intersect at the centre of the maelstrom.

The rest of the solar system was constantly monitored in case the verse-hole 's outlet appeared. Fortunately nothing untoward was seen and, after eleven hours, the particle beams were switched off, leaving just 2% of Venus's carbon dioxide atmosphere behind. Plant life would one day convert this into organic carbon and atmospheric oxygen once living things had been brought to the planet.

After the removal of Venus's heavy atmosphere there was a short but intense increase in volcanism. This was said to be due to removal of the atmospheric pressure which had been plugging many volcanoes. However, the removal of the atmospheric pressure also had the effect of reducing the pressure on magma chambers, which meant that volcanism declined in the longer term.

The plan to increase the speed of Venus's rotation which had been developed in 2633 was initiated in 2680, with the first massive twin asteroid strikes. The impacts were on opposite sides of the planet and were of balanced mass and speed so as to increase angular momentum without disturbing the planet's 225-Earth day solar orbit.

Repeated twin strikes took place at 17 day intervals, which was the time needed for any planetary gyrations to settle. Altogether 48 asteroids impacted the planet over a period of just under 13 Earth months which equated to two Venus orbits, accelerating the planet's clockwise rotation so that a Venusian day would be of similar length to an Earth day or a Martian sol.

The asteroids and Kuiper belt objects contained many elements critical for life. The list included iron, copper, manganese, zinc, chromium, selenium, lithium, cobalt, silicon, boron, calcium and a wide variety of trace elements. All of these had been chosen from the millions of rocks in the asteroid belt as they would be important for life in Venus's future. The planet's rotation was then allowed to settle for 20 Earth years prior to tilt correction.

The intense bombardment of 2680 initiated a volcanic winter which obscured Venus's surface in swirling black dust clouds. Observers watching the planet waited for each rotation of the planet to settle as close as possible to a 24 hour duration as possible before deciding when to make the tilt corrections.

There was a concern that the impacts would trigger a volcanic resurfacing event, but although survey observations found plenty of volcanism, there was no sign of geological plate sinking. As it was, the impacts had all been around the planet's equator; meaning that a ring of fire developed across the Aphrodite Terra highlands and right around the circumference of the planet.

The tilt correcting impacts bore more risk than the rotational enhancers as the danger of inducing a dynamic wobble was greater. However, the sequence, masses, targets and speeds had been recalculated repeatedly and these calculations were corrected as the planet's rotation settled.

By 2700 Venus' rotation had settled at 23 hours and 35 minutes, which was close enough to an Earth 24 hour day to allow tilt correction to take place. Each of the four twin impacts was planned to increase Venus's 3-degree tilt by another 5 degrees, with the target being 23 degrees - similar to Earth's. The first impacts happened on the 27th May close to the north and south poles and the last on the 21st August. The scientists then waited to see what rotation and tilt had been achieved and if any wobble had been induced.

With the experience of terraforming Mars in their minds, there was a great deal of discussion amongst the scientists as to Venus's future climate. Much of the debate focused on ways to keep a hot planet – one which was closer to the sun than Mars or Earth - as cool as possible. The concern was that, if the scientists failed to create the desired climate, the planet might either not be terraformable at all or it would be too hot for humans.

One of the suggestions proposed during the 2633 Venus conference had involved seeding the upper atmosphere with floating plants which would excrete oxygen into an internal float. At the time, this suggestion had not been taken seriously, but in retrospect it was considered to have value as a way of shading the planet.

After the impacts had ceased and the dust had settled, the 2% portion of the planet's carbon dioxide atmosphere was intended as fuel for photosynthesis to convert into organic carbon and atmospheric oxygen. This would be supplemented by the gases produced by the planet's volcanism. However, first the volcanic winter had to end and oxygen, nitrogen and water had to be brought to the planet.

Over the course of the following century Venus's atmosphere cleared, the volcanic winter came to an end, the surface cooled, and the planet's rotation and tilt stabilised themselves. This phase allowed the surface to be investigated in detail for the first time.

2 – GEOLOGY

Over the preceding hundreds of years many rovers and exploratory craft had mapped and analysed the planet Venus's surface. Unfortunately, Venus's superheated atmosphere had rapidly damaged or destroyed the majority of these craft. Despite this, rock, ash and sand samples had been successfully collected and returned to Earth, allowing scientists to confirm the basaltic nature of the majority of the planet's surface.

In fact, Venus was almost completely covered with basalt lava. This was because Venus's lavas were extremely fluid and could flow for hundreds of kilometres under its superheated atmosphere before cooling and solidifying. This had led to the development of pancake-like shield volcanoes which looked from space to be completely circular. There were fewer explosive volcanoes as most of the lava was runny rather than sticky. In spite of this, there was enough volcanic ash to allow the development of dunes, and the addition of dust to the fast-moving atmosphere had been the primary weathering mechanism.

Venus was known to have a 3,000 kilometre iron core overlain by a 3,000 kilometre mantle. There was no known boundary between the crust and the mantle as there was on Earth, meaning the crust could not flow and cause plate tectonics to take place. On Earth, the tectonic structure meant that the plates could float on the mantle and could flow across the planet to collide or slip under each other to create mountain and volcano chains. However the presence of granite in Venus's main continents necessitated the presence of water and plate tectonic subduction in the distant past for their formation. Subduction happens when one plate flows under another and this was the process by which granites were formed on Earth. All of this was taken to possibly mean that the early Venus had had seas and an atmosphere before the greenhouse effect had evaporated all the water from its surface.

Another factor was that there was an absence of meteor impact craters older than 500 million years, which meant that the whole planet may well have resurfaced at about that time. It was also possible that Venus's superheated atmosphere had served as an insulating blanket collecting heat from the surface and from the sun, thereby keeping lavas liquid and flowing, and that over the course of time, the whole surface had been resurfaced gradually rather than all at once.

The huge amounts of volcanism on Venus's surface led to the creation of many thousands of fault lines where compression, extension and lateral forces split the crust into forms not seen elsewhere in the solar system. These included tesserae where the crust was split into tile-like boxes and arachnoids where crater collapse had created fractured landscapes which looked like spider webs. Together these hinted at the surface being highly earthquake-prone which made colonisation both dangerous and difficult. It was also believed that the addition of large amounts of water to Venus's surface would also allow some of it to sink through the crust and restart plate tectonics. For this reason, Venus was expected to become more earthquake-prone rather than less.

The land form seen when people first descended to the surface was one of an angular, sharp, un-weathered nature and included lava tubes of up to 700 kilometres long. Much of the ash originally present on the surface had been removed during atmospheric extraction. However, new deposits had been created by the meteor impacts -which meant there would be places where plants could gain a foothold once an atmosphere had been added. It was also discovered that when some local rocks came into contact with water they shattered, meaning that rain would one day make more of the surface amenable to root development.

After colonisation began a few deaths were caused by sudden carbon dioxide releases which asphyxiated people, leading to the creation of no-go zones on the surface where people and animals were at the greatest risk.

Once the planet had an atmosphere and water, geologists were able to explore the surface and watch the first evidence of water weathering on Venus's surface. There was also a hunt for sediments which might contain ancient fossils from a three billion year old past, but to-date these have not been found.

3 - ATMOSPHERE

The next step in the terraforming process was to create atmosphere and oceans. Therefore, thousands of frozen oxygen and nitrogen objects were brought by tug-bots from the Kuiper belt to be skimmed into the atmosphere. The creation of these tug-bots was by 2750 an autonomous process and their main breeding grounds were certain metal-rich bodies in the asteroid belt. Here a nano-robotic seed was planted on the asteroid's surface which allowed the nanos to replicate and build many thousands of copies of a single component. Each asteroid served as a factory for a specific part of the final machine, and they were connected by assembly robots which moved component parts to another asteroid where they were fixed together.

As such, there were asteroids on which components like drives, batteries, motors, shields, funnels and lasers were constructed. Once it had been dedicated to a process, the asteroid was reserved so no other sorts of nano could be produced. This did lead to some problems when evolutionary design systems upgraded components, requiring the complete nano-sterilisation of an asteroid before the new process could commence with new nanos.

During the rotation and tilt correction process, Venus's mass had been increased to 85% of Earth's which meant the atmospheres of the two planets were likely to be quite similar. However finding these key resources beyond the gas and ice lines in the Kuiper belt was difficult and had taken a long time. This resulted in autonomous survey craft being sent out to locate the necessary objects. In the first generation of these craft, the information about an object's size, speed, direction and composition were beamed back to a base station, which triggered a tug-bot to collect the resource and take it to Venus. The second and subsequent versions of these craft included multiple tug-bots in the superstructure of the survey craft. This saved time, as once an object containing the required minerals or elements was located, a tug-bot could be released to take it directly to its destination.

Debates took place as to whether Venus should have a moon, but opinion amongst mathematicians, astrophysics and planetary scientists was divided. Some thought a moon would stabilize Venus's wobble and make the planet more amenable to life, whilst others believed it had been stable before terraforming and so there was no reason for the planet to have one now. This is a debate that continues –but, even today over two hundred years later, moving moons safely is not something we can easily do.

By 2830 atmospheric pressure on Venus's surface was sufficient to allow the next phase of the terraforming process to commence. For this phase water ice was needed to create oceans, lakes, rivers and clouds. To facilitate this, huge lumps of the correct type of water-ice were brought from the Asteroid and Kuiper Belts to make Venus life-ready. These were skimmed into the top of the atmosphere in their thousands, creating a rainstorm which lasted more than ten years. Huge torrents tore their way down from the volcanic highlands into Venus's basins, ripping their way through lava flows, ash fields and volcanic craters. Maps of Venus's topography which had been drafted prior to the impacts had to be redrawn with oceans and rivers. The positions of the highlands changed as the impacts had added new mountains to the planet's surface. Most of these new mountain chains were around the equator, though two new mountainous regions had appeared close to both North and South poles.

Because of Venus's mass, faster rotation, shorter years and the amount of water being delivered to the upper atmosphere, weather systems soon became a major factor in the terraforming process. A stratosphere became evident and began developing ozone and ionosphere layers which offered some protection from the solar wind. The auroras were however very poor due to the planet's weak magnetic field and these layers gave little protection from solar and cosmic radiation. This reinforced the fact that an artificial magnetic shield would be needed for Venus, much like the one which had been developed for Mars hundreds of years before. This was put into place in 2933 with 49 fusion-powered magnetic field generators.

Because Venus was just 108 million kilometres from the sun its atmosphere received far more solar energy than Earth's. As a result thousands of machines were built to create a heat shield in Venus's upper atmosphere. These comprised self-repairing reflective balloons which were filled with helium harvested in the upper atmosphere. They were more effective than the floating life forms which had been seeded in 2894 as they were more resistant to sun damage and tears, and could harvest helium which the life forms could not do.

4 - SEEDING LIFE

Once the planet had sufficient water on its surface, thoughts began to turn to introducing life forms. The terraformers were able to look back at all of the lessons which had been learnt hundreds of years before on Mars and draw conclusions from the poor rate of species survival and the ways in which some species and ecosystems had come to dominate that planet.

The main conclusion reached was that mixing ecosystems generally led to large scale die-offs and that, if two ecosystems were to be seeded, there had to be some form of barrier or gradient between them. This could take the form of a physical barrier like a mountain chain, a temperature barrier like latitude, a chemical barrier like a salinity difference, or a water availability barrier like the edge of a desert. Ensuring that the ecosystems were kept separate was the best way to prevent one from destroying the other. There had to be transition zones where ecosystems met; but, as far as was possible, ecosystem introductions were to be planned within boundaries or gradients so that they were protected from excessive intermixing.

With 500 years of engineering experience, the machines which were used to seed life had changed completely from manmade mechanisms to nano-constructed antonymous devices which could change their form, structure or function quickly to do the job they were required to do. This meant that a seeding shuttle was autonomously assembled from millions of nano-robotic components without human intervention, and could change the species or the seeding method without outside intervention.

The starting point for a seeding shuttle to seed life was a genetic DNA sequence stored in a computer memory. This was printed out by the shuttle, using amino acids to create the needed DNA sequence. This was then injected into an artificial cell or cell nucleus, supplemented with other molecules also constructed from amino acids, and then electrically stimulated. Once functioning, this cell was either cultured until the required amount of growth had been achieved or injected directly into Venus's regolith.

Seeding shuttles floated a metre above the surface and pushed plants or other organisms into the soil using telescopic spikes. In this way thousands of plants or simple organisms could be planted each hour. The terraformers had very little input into this process apart from watching it happen, as their primary roles were experimentation and analysis.

Analysis of the amount of carbon dioxide being produced volcanically led researchers to propose that changes be made to the life lists to allow for super-sized plants to absorb the excess carbon. As Venus was hot enough already, this excess carbon dioxide was definitely not required in the atmosphere.

5 - NANO-ROBOTIC ECOSYSTEMS

Where life forms were more complex than microorganisms and invertebrates, the seeding process either allowed for additional periods of growth and feeding, or a nurturing phase involving a specialised nano mother. These mechanical mothers had been used on Mars in a rudimentary robotic form, but on Venus they were comprised of thousands of nano-robotic parts and could change their form or function in just a few hours.

This often meant that key parts of an ecosystem could be replicated in non-living form prior to the introduction of the animal itself. Thus key ecosystem partners which had an influence on a young animal could be duplicated so that the creature learnt the key survival lessons before it faced the challenges for real. The result of this was that animals which were duplicated in robotic form included predators, clan members, competitors; plus both suitable as well as unsuitable potential prey.

This last category - unsuitable or dangerous prey - worked by delivering sharp electric shocks to any animal attempting to catch them. Whilst not as dangerous as the sting, venom, poison, spines, teeth or claws of the real animal, the electric shocks served to teach young creatures which prey should be avoided. Suitable prey allowed young animals to learn the key lessons involved in finding, stalking and hunting, and once caught would deliver an edible reward in the form of a tasty fluid or gel which would encourage the young animal to repeat the behaviour.

Once a key animal had been introduced in living form, its nano-robotic duplicate was gradually withdrawn. This allowed the rest of the ecosystem the time it needed to learn the correct responses and strategies necessary to ensure survival.

The shuttles which introduced these nano-robotic machines were commonest on the edges of a developing ecosystem, either on the barrier between life and bare lava, or on the boundaries between ecosystems. This allowed whole ecosystems to learn key survival behaviours through repeated exposure prior to the real creatures becoming part of the biome.

These nano shuttles were semi-autonomous and could manufacture and activate nano-robotic animals either as a result of their own analysis of an ecosystem or as a direct result of human instruction. So, a shuttle could, for example, categorise a young animal as needing predator knowledge, create a nano-hawk and introduce it to the prey so that the young animal could learn avoidance strategies. The nano-robotic mother was able to work in harmony with the nano predator to ensure the young creature learnt the correct avoidance techniques such as running, hiding or fighting.

6 - THE POLAR CONTINENTS

By 2840 equatorial summer temperatures were over 60-degrees Celsius and 25-degrees Celsius at the poles. In winter temperatures dropped to 40-degrees at the equator and 10-degrees at the poles. Humidity was almost always high due to the constant rain, but as the decades rolled on this declined. Scientists began to prepare lists of life forms which were expected to be able endure the temperatures on each continent. Life lists and life game strategies like those developed on Mars were built to take Venus's challenging climate into account. In the mid-2840s, massive renewed volcanism was detected accompanied by regular earthquakes and tsunamis.

Once life seeding was underway, a human presence on the planet would be needed to oversee the process. Ishtar Terra, the northern continent, was the most likely place for this colony to be placed, although there were also other possible sites close to both poles. Many parts of the surface continued to be affected by large earthquakes, delaying plans for colonisation.

When Venus's basins had filled and the atmosphere had cleared, it was seen that the planet had six continents; two near the North Pole, one close to the South Pole and three along the equator. Three fifths of the planet was covered in ocean and as atmospheric and oceanic flow patterns stabilised themselves, some idea as to currents and weather patterns could be discerned. On Earth, oceanic currents and weather systems flowed clockwise in the northern hemisphere and anti-clockwise in the southern. However, on Venus, because of its clockwise spin, these flows were reversed.

The absence of a moon and therefore of tides meant that wind and currents should have been slower than on Earth. Nevertheless, the extreme variations in temperature between the equator and the poles drove intense weather systems which meant that cyclones and hurricanes were as common on Venus as they were on Earth.

Because Venus was much hotter than Earth, and due to the amount of moisture and the absence of cold winters, many plants grew constantly rather than just during the spring and summer. There were temperature-related problems for animals living near the equator, meaning that trees could not use animals to perform useful functions like pollination or seed dispersal without their being given heat-tolerant attributes by humans.

Venus had three continents which were good for life in the forms which had existed on Earth, and three continents where life needed to be modified to survive the intense summer heat. It was therefore decided that the three polar continents were to be terraformed first and the equatorial ones later on.

On each of Venus's three polar continents, the first living things to be added to the lava and ash were nitrogen-fixing blue-green algae and cyanobacteria. These formed thin mats and were able to assist in, holding water, extracting nitrogen and stabilising the soils. The introduction of algae and cyanobacteria was followed by that of lichens and ferns, some of which could also extract nitrogen from the air, and some of which could form a ground covering some centimetres thick. The third stage in all regions was made up of herbs, creepers, lupins, myricas and desmodiums; which could enrich nitrogen, stabilise soils, and had small wind-distributed seeds which could be blown hundreds of kilometres.

Once these stages had become established, ash and lava were ready for the next stages, including the introduction of soil microorganisms and trees.

6.1 - ISHTAR TERRA

The source of life for Venus's northern Ishtarian continent came from the island of Madagascar off the east coast of Africa. It was a large exaggeration to call Madagascar "a continent" as it was merely 1,500 kilometres long, whereas Ishtar was 5,600 kilometres across. However, both islands were subdivided by a central mountain chain which caused rain to fall on one side of the island and dry desert conditions to predominate on the other. Madagascar's mountains were up to 2,900 metres tall and Ishtar's were over 11,000, but the overall effect of climate division was the same in each case.

Madagascar's rainy season lasted from December to March during which time the eastern and northern rain forests were subject to tropical cyclones and heavy rainfall, with average temperatures in the region of 30-degrees Celsius. The eastern side of Ishtar Terra had similar temperatures and was affected by gales which brought intense rainfall for sixty days out of Venus's 225-day year.

So, whilst Ishtar was considerably larger than Madagascar, the distribution of its wet and dry regions was similar. Additionally Madagascar had 9 koppen climate zones and 12,000 species found nowhere else, and so had enough life and was diverse enough to seed a continent much larger. This meant that each part of Ishtar could be classified by its climate type - and then linked to a similar region on Madagascar, from which a transplant ecosystem in DNA form should be sourced.

The rocks which made up the eastern two thirds of Madagascar were either granite or gneiss, while the surface of Venus was almost completely composed of basaltic volcanic rocks. There were differences between the chemistry of granite, gneiss and basalt meaning that Venus's rocks were richer in calcium, iron and magnesium, but poorer in silica. Nevertheless the chemistry of granites, gneisses and basalts were similar enough to mean that life on Venus would be able to find the minerals it needed to survive.

One major problem which faced the terraformers was that many Madagascan forms of life had become extinct due to human pressures and changes in climate. In consequence of this, the terraformers had to look in other places for the critical DNA they needed to resuscitate parts of some damaged ecosystems. Some species had been conserved in living form in zoos; others had been kept as dried, frozen and preserved tissue samples; - and others had had their DNA sequenced and kept in data form. Over the centuries Madagascar's population had expanded from under 5 million to a peak of over 35 and then declined again to under 15 million. This had put great pressure on the island's wildlife and some of the island's most charismatic species had vanished.

The nine zones found on Madagascar could be broadly summarised as three ecoregions: tropical moist forest, hot dry forest and wetlands.

6.1.1 - TROPICAL MOIST FOREST

The eastern fringe of Ishtar Terra, like the eastern edge of Madagascar, did not have a dry season. This meant that the strip nearest to the eastern ocean had daily rainstorms throughout the year and this was the primary reason for this portion of Ishtar being the first place on Venus to be seeded with life.

Unlike Mars where samples of microorganisms were brought in frozen form from Earth, both technology and methodology had advanced to such an extent that this was no longer necessary. Instead all life forms including microorganisms could be printed from their DNA data when and where they were needed and then planted directly onto Venus's surface. This also applied to nuts and seeds which were of notable importance as, during the terraforming of Mars, nuts and seeds had been sourced from Earth and had consequently introduced a number of unwanted tree diseases to Mars; something those involved in the terraforming of Venus were keen to avoid.

By the 2890s, new generations of smart seeding machines could print and plant new bacteria in just a few seconds, and even larger - more complex animals could have embryos created in minutes. The functions of the thousands of microorganisms which make up soil had also been discovered which meant that the most helpful forms could be seeded to perform specific jobs such as accessing key minerals and nutrients. The advances in DNA printing, artificial intelligence and shuttle engines meant that very little of this process needed to be done by humans. Once designers had chosen the species to be introduced, selected the region for the introduction and set the start date, the whole process was completed by machines without human interference. This included the seeding of species which were dependant on other species, so, once a tree had been seeded, the hundreds of fungi, insects and microorganisms which depended on it for survival could be gradually added as well.

The tropical moist forests of Madagascar's eastern seaboard were more complex than even the Koppen climate zone classifications suggested. In fact, there were six different forest ecosystems under the one "moist forest" classification which included coastal, lowland, montane, cloud, high elevation and sambirano forest. The higher and further west the forest, the longer was the dry season and therefore the more critical was the timing of planting at the beginning of the rainy season.

The first autonomous shuttles descended to Venus's surface in 2894 to begin the seeding process. As with Mars it was soil micro-organisms, nematodes and arthropods which were added first. The timing of introductions was determined by which species did well and when conditions were right, so ecosystems were built incrementally. The ships which began the life seeding process were mostly small, as was the equipment which printed micro-organisms. Later stages in the life printing process created more advanced multicellular organisms and so needed larger printers and bigger ships.

The microorganisms chosen to be introduced first were selected because of their ability to access key resources. Hence bacteria which could extract nitrogen from the air were prioritised, as were organisms which could enrich phosphorus or potassium or other key elements. These same criteria also applied to the first lichens and plants, with making the soil suitable for growing forest trees as the goal. Not all of the primary microorganisms, lichens or plants were from Madagascar as the ability of each of these life forms to enrich the soil was considered more important than sticking strictly to Madagascan life forms. The first microorganisms were planted over a metre beneath Venus's surface to avoid desiccation, and the first plants were located in spots where they would be shaded from the sun's intensity.

On Ishtar's eastern fringe with its daily rainfall, plant desiccation was not an issue. However, further inland planting sites had to be chosen with some care. A shading cliff or hill was an advantage to the first plants as, until their roots had penetrated the hard-packed Venusian regolith, lack of water would kill even the toughest of species. Where lava flows made the surface resistant to planting and root development, pockets could be filled with soil transported from other sites. The deeper crevices had the advantage of providing plants with a readymade all-round sunshade, increasing the chances that their roots would remain hydrated and that their leaves could reach the bright sun after a metre or two of upward growth.

After seeding with microorganisms, it was 20 years before the first plants could be added. This was time enough for the regolith to change into soil as mineral enrichment changed its composition and invertebrates changed its structure. It was quickly noticed that, with high temperatures and constant supplies of water, organisms grew very quickly. This meant that high temperatures and water together were able to chemically leach some nutrients out of solid rock, the impact of which was that life was not as reliant on artificially added minerals and nutrients as had been the case on Mars.

The first plants were planted bordering the sea, lagoons and ponds, and were salt tolerant Madagascan varieties which were highly efficient at extracting nitrogen from the air and nutrients from the regolith. Most had nodules on their roots containing cyanobacteria which could collect nitrogen and phosphorus. Without these key soil elements, plants would not have grown.

The first true trees to be planted included the salt tolerant mangrove species which would later serve as nurseries for juvenile fish, crustaceans and invertebrates. Lowland rainforest trees were the next to be planted, and the sites chosen had to be both warm and very humid. They could be introduced at altitudes of up to 1,200 metres and preferred to keep their roots out of direct sunlight. At lowland sites where the air was less moist, sambirano forest trees were added. Higher up, at altitudes of up to 2,000 metres, montane and cloud rainforest trees were seeded; and above these were planted the high mountain varieties, all dependants on local humidity and temperature.

Because of Venus's high temperatures and thick cloud cover, most mountain species prospered at altitudes 50% higher on Venus than on Earth. At very high elevations above 3,000 metres the seeding started with scrubby small trees and progressed to secondary additions such as mosses, lichens, ferns, grasses, small palms and orchids, all of which did well in the moist cloudy conditions. Secondary plant seeding at lower altitudes included ferns, bamboos, vines, orchids, mosses and lichens.

Stage three seeding in the 2930s included many species of pollinating insects such as bees, butterflies and moths; as well as ants, beetles and flies. Later spiders, dragonflies and mantids were introduced to keep insect populations controlled and healthy. These and the seed and fruit eaters introduced a decade later helped spread forests across Ishtar's eastern seaboard. By 2960 Ishtar's eastern rain forests contained many species of lemurs, chameleons, snakes, bats, crabs, civets, frogs, tenrecs, skinks, iguanids, geckos and birds. All of these helped the forests to expand rapidly along the east coast, so that in the 2970s predators including eagles, owls, crocodiles, turtles, monogeneses and fossas could be introduced.

Attempts to bring giant lemurs back from extinction failed as the few bone and tooth samples which had been discovered contained DNA which was too damaged to allow the extraction of a full DNA sequence. However, it was believed that they had been important forest seed dispersers, so efforts continue. The last introductions to take place in rivers and lakes within the rainforest biome were two species of Madagascan hippo. These were brought back from extinction with DNA from living African pigmy hippos and the preserved bones and teeth of the two extinct Madagascan species.

6.1.2 - HOT DRY FOREST

On Madagascar hot, dry forests occurred in the south and west of the island. They included dry deciduous forest, limestone tsingy forests, tapia woodlands and spiny forest. Those which could be seeded on Ishtar had to be supplied with sufficient water and nutrients to allow them to become established. This was, for the most part, done by the same microorganisms, lichens and colonising plants which had been used on the eastern side of Ishtar.

The most problematic of these ecosystems was the limestone tsingy which was composed of needle sharp rock pinnacles. All of the plants and many of the animals relied on the presence of lime-rich rock which was absent on Venus apart from in places where calcium-rich asteroids impacted the surface near the equator and, to some extent, where it was present as a constituent mineral in basalts. Therefore, apart from providing a source of some species not present elsewhere in Madagascar, limestone tsingy could not be duplicated on Venus.

Dry deciduous forests were almost entirely confined to regions of western Madagascar underlain by sedimentary or lime rich-rocks. This meant that, like the tsingy forests, most plants from this source would not survive on a basaltic substrate on Venus. Nonetheless, a portion of the dry forest ecosystem was on cretaceous volcanic rock, meaning that these areas could serve as a source for species which had a better chance of surviving on the western part of Ishtar. The all-important microorganisms were sourced from this region, as were soil invertebrates and small plants. The larger plants, sourced from a wider area, included baobabs, palms, lianas, and tamarinds, and the animals comprised lemurs, mongooses, tenrecs, reptiles, amphibians, birds and carnivores like the fossa.

Tapia woodlands occurred in the central highlands of Madagascar, on igneous rocks and at altitudes of 500 to 1,800 metres. Over the last thousand years there had been a great deal of degradation of this region but, as Madagascar's human population had fallen, there had been some recovery in the life forms found in the woodlands, but many species were long extinct. The tapia was the dominant tree and its roots were swathed in ectomycorrhizal fungi which helped the tree collect water, nitrogen and phosphorus. The shrub layer included acid loving shrubs, grasses, and lianas, all of which preferred igneous to lime-rich rocks. This made them ideal for introduction to the dryer mountainous parts of western and central Ishtar.

Animal species which still existed in this ecosystem included lemurs, shrews, tenrecs, rodents, birds, reptiles and amphibians. All of these remaining species were seeded onto Venus's growing tapia woodlands; however, Madagascar's central highland region was once populated by a number of elephant birds, a giant tortoise, and giant lemurs. The giant lemurs as previously mentioned were problematic, but Aldabra giant tortoises, being related to the extinct Madagascan form were seeded onto Ishtar in the 2980s and the first resurrected elephant bird was seeded in 2990.

The spiny forest was an arid ecosystem rich in didiereaceae cactus-like woody plants with small deciduous leaves protected by sharp thorns and spines. They grew almost entirely on sedimentary rocks near the southern tip of Madagascar, apart from some small patches where they extended a short way onto gneisses and granites. Over the centuries these spinney forests had given some protection to the creatures endemic to the area, but there was some loss of such forestry due to cutting for charcoal manufacturing. The patches on igneous rock and on red sandstone served as sources of microorganisms, invertebrates and small colonising plants. Animals endemic to these forests included tortoises, geckos, chameleons, iguanas, tenrecs, lemurs, mongooses and birds, all of which were seeded once spinney forests were established on western Ishtar.

In some areas of Madagascar's central mountains patches of the original high elevation scrub forest remained. High up on the mountain slopes a mixture of stunted montane lichens, peat bogs, grasses, orchids and palm trees survived. Although they had been badly degraded by fire, cattle browsing and climate change, those patches which did survive were very important for the considerably taller volcanic peaks of Ishtar Terra. Thus from these small patches came the life to colonise thousands of square kilometres of central Ishtar.

6.1.3 - WETLANDS

There were a number of separate ecosystems in Madagascar which fitted under the loose description of "wetlands". These included mangroves, lagoons, lakes, rivers, marshes and swamps which supported a large proportion of Madagascar's wildlife.

Madagascar had large areas of mangrove forests on its eastern fringe. They were typically found in silty, saline locations such as rivers, estuaries and muddy coastal regions, and plants were able to survive in these environments because they could extract oxygen from the air using breathing roots which emerged from the mud around the tree. Mangrove forests served as a critical habitat for both birds nesting in their branches and for fish living amongst their roots.

The locations chosen for the first seeding of Mangroves were selected with some care on Ishtar. This was because wave action was expected to alter the form of the coast as sea and wind currents became established. For that reason, the first locations chosen for mangrove seeding were newly-formed estuaries at the mouths of rivers where large-scale silt build-up was predicted to be permanent. Without tides the only way mangroves could survive was by cyclic wetting and drying by wave action. This meant that the mangroves were less successful on Venus than on Earth and were only able to establish in regions affected by frequent storms.

The key reason for mangroves being seeded as a part of the tropical moist forest seeding programme was that they served as nurseries for the young of marine species. The lack of tides and the thin mangrove band around Ishtar meant that terrestrial forests were able to colonise areas much closer to the sea than would have been the case on Earth. Mangroves then occupied a thin coastal strip fringing coasts, estuaries and deltas. They would be a much more important ecosystem as part of the terraforming of another planet, but one with a moon and tides. However it is possible that Venus may be given its own moon in the future to stabilise a planetary wobble or to improve its biodiversity.

Away from the coast, Madagascar's Inland lakes, rivers and swamps were as diverse as its rainforests. With an elevated plateau, short, fast-flowing rivers crossed granite and gneiss to the east, and slower rivers traversed large areas of sedimentary basement to the west. Due to climate and geology, freshwater ecosystems could be split into eastern and western, and it was those eastern biomes which were more amenable to transplantation on Venus.

On Ishtar the eastern highlands were wet and fed many rapid flowing rivers which tore their way to the sea through mountains, craters and canyons. These were seeded with life from Madagascar's Nosivolo river ecosystem as the rock substrate was compatible with that found on Venus. The Nosivolo was the the most important river in Madagascar in terms of biodiversity as its remoteness had protected the region's wildlife after much of the rest of the island had been deforested and farmed.

The first species to be seeded into Ishtar's rivers were microorganisms which were able to access nutrients and oxygen in the fast-flowing currents and to feed species higher up the food chain. Secondary seeding was mainly composed of aquatic plants which helped to provide food and oxygen for the next stage which included invertebrates such as aquatic insects, shrimps, crabs, crayfishes and gastropods.

Because of the precise nature of the DNA printing and seeding, invasive water ferns and water hyacinths which had clogged many of Madagascar's waterways could be excluded from the seeding on Ishtar.

The fourth stage introductions included many species of freshwater fish including Madagascan catfish, cychlids, goby, killies and silversides, and the final stage introductions included amphibians such as frogs, reptiles like turtles and lizards, and mammals including the aquatic tenrec.

6.2 - TETHUS TERRA

Venus's second northern continent, created by the asteroid impacts which corrected Venus's tilt was smaller and less mountainous than Ishtar Terra but no less affected by easterly winds. Average summer temperatures of about 30-degrees Celsius on Ishtar and Tethus were similar due to their comparable latitudes. They were also close to the temperatures experienced in the regions near Earth's equator from which life needed to be sourced.

Because of the accuracy of the asteroid impacts this second continent was shaped like an elongated doughnut, with high fringing mountains and lower central ranges of hills, valleys and lakes. At 1,800 kilometres long and 1,000 kilometres wide it provided wet humid habitats on the fringes and dryer conditions in the centre.

The source of life for Tethus Terra came from the African rift valley as both locations comprised a roughly elliptical region of fringing mountains and a central flat area with hills, plains and lakes. The eastern ecosystem of the Udzungwa Mountains in Tanzania was transplanted to Tethus's eastern mountains and the western Virunga ecosystem from Rwanda and Congo was transplanted to Tethus's western mountains.

In this way life on Tethus Terra was able to take advantage of both Venus's weather and rainfall patterns and of the diversity of life in Africa's rift valley.

6.2.1 - UDZUNGWA MOUNTAINS

The east African Udzungwa rain forests were a mountainous region in southern Tanzania. Its elevation and isolation meant that Udzungwa was home to many plants and animals that had evolved in such a way that they were unique to that region and could be found nowhere else.

The Udzungwa Mountains were underlain by Precambrian gneisses, very much like eastern Madagascar, and rose spectacularly out of lowland plains to heights of over 2,500 metres. Rainfall of up to 3 metres per year, blown in from the Indian Ocean to the east, made the region highly diverse with more than 2,000 plant and 300 animal species. There had been species extinctions in the previous few hundred years, mainly due to climate change and local human population pressures, but as Africa's populace declined after 2200, cool mountains and national park status provided a refuge for wildlife from the heat of the plains.

The introduction of soil microorganisms worked very well on Tethus due to its chemical similarity to the Eastern Highlands. With as much rainfall in eastern Tethus as in the Udzungwa Mountains, regolith became soil relatively quickly. Equally, the first plants liked the wet warm conditions and, in the absence of competition, grew rapidly when the right nutrients were available. Once semi-mature woodland had established itself, bryophytes, mosses and lichens were added to create habitats in the upper canopy. Later grasses and bamboos were seeded to colonise those areas where trees had failed to thrive.

Third stage introductions included many species of insects including beetles, ants, mantids, butterflies and moths; as well as spiders, millipedes and centipedes. Following this stage of introductions, 40 species of frogs, toads and small insectivorous or herbivorous mammals were introduced in stage four and 35 species of reptiles including chameleons, snakes, geckos and skinks in stage five. Both amphibian and reptile species had suffered extinctions as the Earth's climate had warmed and species which preferred cooler conditions had been forced up the mountains until they ran out of habitat. Fortunately, samples of all of these species had been kept in dried, preserved or sequenced DNA data form and were able to be recreated, albeit sometimes in forms which were subtly different from the original species.

Stage six included the introduction of browsers and grazers such as monkeys, birds, antelope, mongooses, hyrax, pigs, aardvarks and porcupines, and stage seven comprised predators like leopards, owls and honey badgers.

6.2.2 - VIRUNGA VOLCANOES

On the west side of the rift valley the Virunga volcanic terrain ecosystem spanned the boarders of Rwanda, Congo and Uganda, and provided a chemically-suited collection of life forms for DNA transplantation to Tethus's western mountains. The prevailing winds and hence rainfall were again from the Indian ocean to the east and so the most diverse parts of the ecosystem were on the eastern side of this volcanic range.

The rainfall pattern in Virunga was one of twice yearly dry and wet seasons, which though not identical to Tethus's weather system, was similar enough to avoid extended dry or wet periods which might have resulted in large numbers of species becoming extinct. The geology of the two mountain ranges was even more similar than either Madagascar or Udzungwa to Tethus as the rocks were recently erupted volcanic basaltic lavas. Elsewhere in this part of the Albertine Rift, the geology was composed of Precambrian basement gneisses and recent volcanoes, all derived from the development of Africa's Great Rift Valley.

The Virunga volcanoes were a part of the western branch of the African Rift and, although the whole rift covers over 6,000 kilometres north to south, only fragments with highly diverse communities of plants and animals survived past human peak population in the 2100s. As the number of people living in the rift declined and the remaining individuals became wealthier, the appetite for conservation in Africa grew and national parks became larger. This meant that sources of life which could be used to seed Venus were quite widespread. However, like Udzungwa, Virunga remained a biodiversity hotspot and only a few species needed to be sourced from other mountainous parts of the rift valley.

Tethus's western fringing mountains rose to elevations of just under 3,000 metres which was considerably smaller than the 5,100 metres height of some of the Albertine Rift's volcanic peaks. Yet, as the majority of Virunga's diversity lay below the 3,000 meter contour the West African rift remained an excellent source of life for Tethus.

Another factor behind Virunga's suitability was the great variety of habitats to be found in this region. These included forests, lava fields, forests, savannas, river valleys and swamps, leading to the development of a wide range of vegetation types such as alpine, mixed forest, bamboo, open woodland, open grassland and swamp. Each evolved to suit a range of animals and the greater the number of habitats, the more diverse the ecosystem became.

The first life forms to be printed and seeded into Tethus's western fringes were microorganisms evolved to extract resources from lava and atmosphere. Even better matched to Venus's geological chemistry than Madagascan or Udzungwan microorganisms these spread through the moist regolith very rapidly, allowing soil invertebrates to be added later in the same year.

The following year a start was made on adding nearly 1,200 stage 2 plant species to Tethus's freshly created soil. The first introductions were mainly low growing herbs, flowering plants, grasses and mosses. These were good colonising plants, some of which included root nodules containing fungi and bacteria which were able to concentrate key minerals and nutrients. Those which relied on wind for fertilisation and seed distribution spread rapidly around Tethus and appeared in many places where seeding had yet to even begin.

The autonomous seeding machines then searched for evidence of previously added life before adding tree or bamboo seedlings. These were more likely to survive where soil organism diversity had been maximised. So, the presence of microorganisms, invertebrates and small plants indicated that there were sufficient nutrients in place to support the growth of trees. In some places where water had created pools, water-loving plants were added to create swamps and lakes.

Three years later, with plant growth and habitat expansion evident in many places, stage 3 butterflies, moths, flies and beetles were seeded to create a food resource for animals higher up the food chain. The number of arthropod species included in the introduction meant that this stage of the seeding took another three years as successive layers of diversity were added to eastern Tethus. Predatory invertibrates like spiders, dragonflies and mantids were added when other species had had a chance to breed and expand their ranges.

Four years after this, over 100 reptile species including chameleons, snakes, geckos, lizards and skinks; and 70 amphibian species including caecilians, frogs and toads were introduced in stage 4 to control the number of arthropods. This stage, as with the Udzungwa seeding, overlapped with the first small mammal introductions. Therefore rats, squirrels and shrews were added as a food supply for the next stage of ecosystem construction.

Stage 5 introductions included over 250 bird species, 20 types of monkey and various types of antelope, warthogs, forest hogs as well as the first small carnivores such as genets and golden cats. Those birds which had been migratory on Earth had a compulsive desire to migrate, but without a magnetic field they couldn't tell where they should fly. For some years they circumnavigated Tethus but, as other parts of Venus were terraformed, they established migration patterns between Venus's continents. Stage 6 comprised large herbivores including elephants, buffalos, mountain and lowland gorillas, chimpanzees, okapi, hippopotami and giraffes. Finally, stage 7 introductions included lions, hyenas and leopards.

6.2.3 - CENTRAL TANZANIAN WOODLANDS

Sourcing life for Tethus's central plain meant that the DNA of living things from the plains between the Udzungwa and Virunga needed to be sampled. This ecosystem of grass and woodlands extended all the way from Tanzania through Congo and into Zaire.

There were three key areas of biodiversity which needed to be sampled to collect all the important forms of life from this region. They were Gombe in Tanzania, Kafue in Zambia and Upemba in Congo. Additionally, Lake Tanganyika, which boarders Gombe, was sampled for lake life to colonise the lakes which had appeared in the Tethan interior. None of the sources of life for the central region were as diverse as Virunga or Udzungwa, but they were more likely to survive the dryer lowland conditions there.

The underlying geology for all three source areas was Precambrian gneiss so the microorganisms needed to build soil from volcanic regolith on Tethus were widely available. The soils of the region were all nutrient-poor and well-drained, and were only enriched by the action of termites in accumulating plant material. Therefore, only those plants which would grow in nutrient-poor conditions were widespread. Water availability was also poor, as rains fell in the summer months between November and April, and temperatures ranged from 18 to 35-degrees Celsius.

The first place sampled was the Gombe National Park in eastern Tanzania which was originally a small area of high diversity forest and grassland along the eastern shore of Lake Tanganyika. Since 2200 this area had been expanded to include Masito and Ugalla which were originally separate parks. DNA sampling included nearly 700 plant species and the chimpanzees which had been studied for hundreds of years. The area also contained baboons, monkeys, over 200 species of birds, 11 snake species, bush-pigs, pangolins, servals, leopards, elephants, eland, hartebeest and duiker. In addition to this collection of terrestrial DNA, Lake Tanganyika was sampled for the DNA of nearly 300 fish species, snakes, hippos and crocodiles.

The second source of species for the Tethan interior were the Kafue woodlands in Zambia from where more than 3,000 plant species, 500 birds, 30 reptiles, 8 amphibians were sourced. Additionally, elephant, buffalo, black rhino, antelope, hartebeest, reedbuck, lion, cheetah and wild dogs were all species which were sourced from here.

The third DNA source of species was the Upemba Park in the Congo. This park had been badly depleted of large species by war and poaching in the 2100s but, as peace had returned to the region, the missing wildlife had come back or been reintroduced. When the collection of genetic information started there were baboons, zebra, hippos, lions, leopards, elephants, buffalo, antelope, cheetahs, black rhinoceros and many other species present in the area. The papyrus swamps which included the shallow Lake Upemba, with 7 turtles and tortoises, 35 lizards, 4 amphibians, 65 snakes and 2 crocodiles, managed to retain most of its diversity despite the wars and was included in the sampling.

As with the seeding of other types of African life onto Venus, it was the microorganisms which were most critical to terraforming and it was these which were seeded first into the regolith in the centre of the Tethan depression hills and lakes. These and the small plants and lichens which followed were quick to spread across the continent. The seeding of each part of Tethus was timed so as to overlap and spread the same stage at the same time, so there was a lot of mixing even when the distances were large. This meant that a life form's ability to move was critical to translocation. It was therefore the case that plants with windblown seeds spread rapidly, as did lichens and fungi which also relied on the wind. Plants which were reliant on animals to move their seeds were much slower to move unless the animal could travel long distances. Thus some seeds which were spread by birds were able to move right across the continent, and later so were seeds spread by herbivores.

All of this meant that while trees were still gaining a foothold in the mountains, grasses were spreading wherever the wind took their seeds and the centre of the continent was the first place to become fully vegetated.

6.3 - LADA TERRA

Lada Terra, Venus's third and southernmost continent, like Tethys had been created by the asteroid impacts which corrected Venus's tilt. In form it was a 2,000 kilometre long isolated land mass stretching from south of Eistla Terra toward Venus's South Pole. Lada's climate was one of high humidity, 5 metre yearly rainfall and temperatures ranging from 20 to 35-degrees Celsius. Lada's position in the southern hemisphere meant that it was directly in the path of the westerly gales and caught their full force and moisture content like a huge sail. With two ranges of mountains up to 2000 metres high stretching the length of the continent it had dry and wet seasons, but the dry season was short so that high humidity, temperature and rainfall were the norm.

The place on Earth chosen as a source of life for Lada was the in the Osa Peninsula in Costa Rica which had similar climatic conditions. The only major difference was that Lada's mountains, created by the asteroid impacts, were twice the height of Osa's. This meant that while Osa could be used as a source for most of Lada's life, another location had to be found to terraform Lada's higher peaks.

The geology of the Osa Peninsula included large pieces of uplifted ocean floor called Ophiolite which were a good chemical match with Lada's recent lavas and volcanoes. This meant that the microorganisms and invertebrates to begin the terraforming process were common and easy to source.

The Osa Peninsula had been chosen because it had extremely high biodiversity for such a small area. Included were many thousands of species of animals, plants and fungi. Amongst these were 10,000 types of insects, 500 species of trees, 50 amphibians, 50 reptiles, 40 freshwater fish, 400 birds and over 100 mammals. Even in the darkest days of the 21st century when much of the planet had been affected by climate change and peak population, the people of Costa Rica had fought hard to preserve their national parks and Osa was the best preserved of these.

The ecosystems present on the Osa Peninsula included lowland rainforests, upland cloud forests and palm forests, all of which were adapted to high humidity. The collection of DNA from Costa Rican life forms had begun in the 2100s, but much more was done after 2640 when the terraforming of Venus became a real possibility.

The source of life for Lada's mountain peaks was the Arenal Volcano range of recent volcanoes in central Costa Rica. These were primarily cloud forests which helped to supplement the mountain ecosystem by introducing a greater number of cold-tolerant species which could survive at altitude in Lada's mountains.

Once microorganisms had been added to Lada's regolith, the second stage was the introduction of various species of figs. Other trees had been shown to be better at regenerating forests in Costa Rica but these were reliant on much better developed soils, with far more organic material, nutrients and microorganisms. Finding species which would grow quickly on bare lava or regolith with limited supplies of nutrients and poor microorganism development were hard to find. However, figs had been shown to grow effectively on even the poorest soils on Earth and so were introduced as the first trees on Lada.

Fig trees were seeded into regolith where microorganisms were spreading effectively, without adding lichens or small plants first. This was because, although the species of fig trees used were not native to Costa Rica, they were very good at colonising bare rock and excellent at providing food for the animal species to be introduced in later stages. Fig trees had also been shown to be able to colonise almost any location, could produce figs at any time of year, and could release thousands of seeds to further regenerate forests.

Fig trees depended on wasps to fertilise their flowers. These wasps were species specific and would only fertilise the flowers of a single type of fig tree. Figs were also fast-growing and could produce fruit within a few years of planting. In the primary stage seedlings, figs seedlings were added directly to regolith with access to nutrients, microorganisms and water. Once these had begun to grow and to produce fruit, they were able to serve as hot spots of biodiversity for species seeded in later stages with thousands of insects, mammals, birds and reptiles relying on them for food. In the places where figs had been planted, the fig trees' root growth had turned regolith into soil faster than anywhere else.

Once a fig tree was growing successfully, other tree species could be added to the area in stage 3 introductions. Then, when a fig tree had begun producing fruit, the correct types of female fig wasp had to be created, fertilised, dusted with pollen from the same fig tree species before being released close to the tree. Figs were not really fruit at all but collections of inward facing flowers with pollen and flowers themselves hidden inside the fruit. When the fig wasp had found a fig, she bored her way into it, fertilized the flowers with the pollen she brought with her, laid her eggs and died.

When the next generation of wasps hatched they mated, bored their way out of the fig, and the females then went hunting for new figs to repeat the cycle. The fig continued the process by ripening its seeds and awaiting consumption by animals which would do the distribution.

In this way Lada was forested, with other tree species being added as fig trees converted each patch of regolith into true soil. After a zone had been successfully forested, understory and epiphytic plants were introduced to further prepare the forest for animals.

Thereafter thousands of varieties of insects were introduced in successive phases to provide a food source for amphibians, reptiles, birds and mammals. Each introduction was dependant on the proliferation of its food source before it could be added. Thus spiders could not be introduced until the correct types of fly were present in sufficient numbers to supply the spiders with food. The same rule was true of leaf and nectar eating species which needed their key food plants to be sufficiently plentiful to ensure they had a sustainable food supply. The invertebrate stage introductions included thousands of species of butterflies, moths, beetles, flies, cicadas, grasshoppers, ants, scorpions, millipedes, centipedes, mosquitoes, spiders, dragonflies and mantids.

Two years after the first invertebrate introductions, the forest patches had spread and included not just the easy to seed plains and valleys, but also mountain slopes as well. There were still large areas of bare regolith, but the forests had closed the gaps so that 15% of Lada was forested. When the numbers of invertebrates in the forests hit one thousand per ten square metres the first insectivorous amphibians were introduced to damp parts of these forests as stage 5. This stage included the introduction of over 40 species of frogs including rain, glass, and poison arrow species, so that five years later with more patches of forest starting to join up, the numbers of amphibians reached another critical number prompting the introduction of the first reptiles. Fish were also introduced during stage 5 to rivers and ponds. Here they ate aquatic plants and insects which had been added earlier in the terraforming process.

Stage 6 consisted of reptiles, birds and mammals which preferred to eat vegetation, insects or amphibians. These included reptiles like geckos, turtles and small lizards; birds like hummingbirds and parrots; and mammals like tapirs, monkeys, peccaries, anteaters, sloths, bats and agoutis.

After these introductions the expansion of Lada's forests accelerated dramatically. This was because many of the species which had been introduced liked to eat figs, meaning that they were able to spread fig seeds to even the most remote parts of the Ladan continent. Both birds and bats were able to take seeds very long distances, dropping them with a fertilizing packet of guano in even the remotest of Lada's mountains.

Stage 7, again being reactive to species survival and breeding success included reptiles like crocodiles, caimans and snakes; birds like owls and eagles, and mammals like coatis, otters, raccoons, jaguars, ocelots, jaguarondi and pumas.

Stage 8 life forms were not added until the seas had been seeded and included crabs, shrimps, lobsters, manatees and four species of sea turtle.

7 - THE EQUATORIAL CONTINENTS

Venus's equatorial continents were far more difficult to terraform than the polar land masses. With summer mid-day temperatures of up to 60-degrees Celsius, much of the equatorial continental region was too hot and humid for humans and only some animals and plants could endure it. This meant that, whilst humans could visit the equator in winter when temperatures were closer to 30-degrees Celsius, they had to migrate before the summer heat made these regions harmful to their health.

For animals, the problem was very nearly as bad. Some animals such as reptiles were able to endure very high temperatures provided they could also find ways to lose heat. The problem was worse for mammals and birds, which had to migrate to stay alive. High humidity made the problem worse by limiting humans' ability to lose heat by sweating. On Earth, temperatures of up to 50-degrees Celsius were possible, but people had either migrated away from these regions hundreds of years before or else lived underground.

In Earth's tropical rain forests temperatures were typically between 20 and 35-degrees Celsius. The principal mechanism plants used in response to high temperatures was transpiration, which meant drawing in water through the roots, transporting it up the trunk and branches before evaporating it from the leaves. This should have been effective provided there was enough water around the roots to keep the process working. If water was not sufficiently available in the soil, the plant might close its leaf pores to prevent water loss, thereby reducing carbon dioxide intake and hindering photosynthesis. Closed leaf pores could also increase plant temperatures. Hence in order for plants to grow in high temperature environments there had to be sufficient water available in the ground for transpiration.

Trees could survive in these temperatures provided that there was enough water to allow them to lose heat by transpiration. Those plants which were able to grow in hotter, dryer conditions reduced transpiration by dropping their leaves. High temperatures could prevent the correct development of many parts of a plant, making it more prone to disease and unable to produce seeds.

Figs were the primary plants used to begin the creation of equatorial forests on Venus. These had to have their genetic code modified in a number of ways so as to increase their heat tolerance. This meant that certain parts of the fig's DNA had to be swapped for genetic sequences evolved in heat loving life forms which grew in geologically hot places like the pacific-rim volcanic ring of fire, Italy, Yellowstone, New Zealand and Iceland.

These geothermal regions could have ground and water temperatures of up to 120-degrees Celsius, meaning that any organism; plant, bacteria, fungi, virus or animal; had to evolve biochemically in order that it could function in extreme heat. The organisms had therefore evolved ways of producing hormones, enzymes and a large number of other chemicals to cope with their hot environment.

Some of the most useful sources of key heat-loving DNA sequences were plants such as the kanuka from New Zealand, as well as algae, mosses and liverworts from a wide variety of geologically active regions. However, individual DNA sequences which could produce heat-tolerant proteins came from a wider variety of different organisms and locations. This method of replacing normally evolved DNA with the genetic code from heat-tolerant life forms was also used later for many other plants and animals before their introduction to Venus's equatorial regions.

Figs were planted across each of two equatorial continents and on outlying islands by seeding shuttles which could print their heat-tolerant DNA from computer records and place seeds in sun shaded places where water was freely available. Many fig species were used and most were planted straight into damp ground, which would not have worked on Earth where competition with other plants would have stopped them growing. However, on Venus, with no competition, figs were able to quickly establish anywhere provided that there was enough water, their mycorrhizal fungi were present to help them extract key elements and minerals, and there was not too much direct sun or too high a temperature.

The next step in making figs suitable as a pioneering species was to engineer the DNA of fig wasps to allow them to play their critical role in transporting fig pollen from tree to tree. The wasps chosen as being most suitable for Venus were nocturnal species which were able to move pollen between trees in the relative cool of the night. The associations between fig trees and fig wasps were ones in which only one type of fig wasp was able to fertilise each fig species. So, when a fig tree did not partner with a night flying fig wasp, the DNA of both the trees and wasps had to be modified to create a new relationship.

Once figs were growing successfully they could expand using aerial roots to create their own patches of forest. These then served as shelter and food sources for animals, and the same animals in return transported the seeds of other types of trees to create truly diverse forests where only lava and ash had existed before.

Two of Venus's three equatorial continents had their own specific Earth ecosystem source for life. The Amazon was used as a life source for Aphrodite Terra; the Bornean, Indonesian and New Guinea rain forests were used as sources for Devana Terra. However, Eistla Terra was more complex as it possessed no cooler peninsula or island, meaning that it had to be terraformed as an environment in which humans could not live.

7.1 - APHRODITE TERRA

Venus's biggest continent was Aphrodite Terra, which at over 7,000 kilometres long and 4,000 wide was Venus's largest landmass. It was segmented by three mountainous regions; Ouda, Thetis and Mons; and included three great gorges which would one day drain the mountain peaks.

The amazon rain forest was chosen as a source of life for Aphrodite as South America included a large area of highly diverse forest, mountains and rivers. In extent it covered an area nearly 4,000 kilometres long north to south and another 4,000 east to west, and included life from rain forests as far apart as Ecuador in the north, Uruguay in the south, Bolivia in the west and the Atlantic Ocean in the east.

The first parts of Aphrodite Terra to have life added to their surfaces were the two peninsulas at the eastern end of this equatorial continent. To the north the Ulfrun peninsula extended three thousand kilometres north of the equator and to the south the Imdr peninsula spanned about two thousand. This allowed life to be added to regions with average summer day time temperatures closer to 35-degrees Celsius than the more extreme 60 near the equator.

Figs were the first colonising tree species to be planted on Aphrodite, but seeding with heat-tolerant modified life forms such as cyanobacteria, blue-green algae, lichens and ferns happened at more or less the same time. The species chosen had to be good at enriching the regolith with key resources like nitrogen and phosphorus as well as retaining moisture and stabilising ash and gravels. Additionally, introducing species with small wind distributed seeds allowed life to spread rapidly, even without animal assistance.

The nature of the substrate into which a plant was placed was critical to its success. Hence gravel and ash were far better for growing plants than solid lava. The stability of the regolith was also important as ash or gravel which was moving could quickly overwhelm and kill a seedling.

The very first species to be introduced were microorganisms and invertebrates from the Brazilian Mesozoic volcanics and invertebrates from the same source were added to continue turning ash, scree and lava into soil. This was followed by the introduction of colonising plants. Once these pioneering plants were beginning to become established on the Ulfrun peninsula, life from Brazil's Tumucumaque Mountains National Park was seeded to continue the terraforming process. On the Imdr peninsula, life from Bolivia's Isiboro Sécure National Park was used.

Following this a number of additional pioneer plant species were introduced on the Aphrodite peninsulas to make the environment ready for Amazonian tree species. These included lupins, desmodiums and myrica which were all excellent at colonising lava and ash, and at enriching the soil with nitrogen. So, by the time the early coloniser figs were producing fruit and fig wasps had been introduced, the time was right to introduce the first seed dispersal animals.

Fruit bats were chosen as the primary seed dispersers as the species chosen flew at night which meant they could cover more of Aphrodite Terra than birds which were mostly active during the day. Fruit bats were also able to sleep the day through in any convenient overhand or lava tube, so they didn't need to return to the safety of the embryonic forest each day. However, they needed to have changes made to their DNA so as to increase the surface area of their lungs and to increase the oxygen carrying capacity of their red blood cells. These helped them to carry more oxygen and to cool down by losing heat through water carried away during respiration. Additional to these changes modifications were made to their hormones and enzymes allowing them to function normally at higher than usual temperatures.

Another problem the bats faced was that on Earth they normally flew only on moonlit nights when they could see to fly. Venus had no moon and therefore had no moonlit nights, so another way had to be found for the bats to find fruit, trees and lava tubes. The method chosen was to add luminosity to the DNA of both figs and bats. Luminous bacteria were also added to the walls of lava tubes so that the bats could find their way back at the end of each night.

This luminosity was also used in the majority of plants and animals on Venus's equatorial continents. Only predators were excepted so that they could catch their prey without being seen first.

When patches of fig forest were seen gradually encroaching on the central equatorial mass of Aphrodite Terra, the terraformers began to think about introducing more Amazonian species. In the Amazon basin there are estimated to be about 40,000 plants, 400 mammals, 1,200 birds, 350 reptiles, 3,000 fish, more than 100,000 invertebrates and many millions of microorganisms.

The next stages of introductions were additional tree species, introduced to ensure the fruit bats had plenty of food to eat and seeds to spread. When these were producing enough fruit, birds were introduced to help with the seed dispersal. These were the first animals to be active during daytime and the modifications made to their DNA to make them heat-tolerant allowed them to spread seed along the edges of the expanding jungle.

As time passed, bare patches on the peninsulas became hard to find as Amazonian life continued to terraform the two peninsulas. This resulted in secondary seeding being undertaken on other parts of Aphrodite Terra; namely the southern Artemis region which was seeded with life from the Madidi National Park in Bolivia and the northern Bell and Tellus islands with life from Yasuní National Park in Ecuador. Whilst Bell and Tellus were not connected to Aphrodite by land bridges, they were close enough to allow birds and bats to fly between them and to spread seeds to the mainland. Additionally both Madidi and Yasuní were still highly biodiverse which allowed the terraformers access to huge libraries of wildlife diversity.

Tertiary seeding involved invertebrates such as spiders, scorpions, beetles, butterflies, moths, dragonflies and many other types. These provided a rapidly growing food resource for the next stages of animal introductions, which included birds, mammals, reptiles and amphibians. The bird introductions included megapodes which incubated their eggs by burying them in warm volcanic ash rather than sitting on them as was the norm with most birds.

The modifications made to these animals' DNA and their style of life determined how close they could get to the equator and how far they could spread across Aphrodite. Many of these creatures were able to assist with seed dispersal which meant that each new introduction accelerated the rate at which the forests expanded.

The final stage of the process was the seeding of predators into Aphrodite's forests. These included mammals like jaguars, reptiles like anacondas, and birds like the harpy eagle. An attempt was also made to resuscitate from extinction the giant sloth and its predator the sabre-toothed Smilodon, but temperatures were too high for both. However, as the land was terraformed, attention had also been focused on the rivers which were carving their watercourses through Aphrodite's three giant canyons. These included Diana Chasma which drained the Thetis Mountains, Dali Chasma which drained the Mons Mountains and Artemis Chasma which extended the Dali's lowland drainage.

In their initial forms these rivers formed steep, rapidly moving mountain torrents. Yet as they carved their watercourses through virgin lava and ancient canyons, their rapids became fewer and their streambeds deeper. This allowed the introduction of freshwater microorganisms followed by insects, worms, leeches and snails. Later fish and frogs were introduced, with many spreading to forest streams and pools and still later predators like crocodiles, caiman, otters and turtles were added as well.

Lastly the rivers were seeded with Amazon River dolphin brought back from extinction. However the fragmentary preserved remains and poorly stored genetic data did not provide good quality DNA, which meant that despite repeated attempts river dolphins have yet to breed in Aphrodite's rivers.

7.2 - DEVANA TERRA

This continent was as much an archipelago as it was a continent, as the effect the impacts had had on the previously existing landscape was to enlarge the existing upland areas and to create numerous islands, reefs and peninsulas. The primary change affected by Venus's spin correcting meteor impacts was to connect the mountainous regions of Beta, Asteria and Phoebe to create a single 3,000 kilometre central landmass spanning the equator from north to south. Additionally a new peninsula of land was created which extended 2,000 kilometres to the west from the centre of this new continent. Together these made the continent look like a letter T on its side. Additionally the north south bar of this continent was further split by a deep canyon called the Devana Chasma which divided the mountain chain in two, leaving a deep and wide valley between the two upland areas.

Because of the number of islands, including Themis Regio 1000 kilometres to the south, and the way ocean currents were impeded and channelled by the north south nature of the continent, Davana Terra had some resemblance to the islands of the South China Seas. For this reason it was life from Borneo, Indonesia, Java, Sumatra, Sulawesi and New Guinea which were used to seed this new continent.

During Earth's ice ages, many of these islands had been connected by land bridges, and thus they represented separated parts of the same ecosystem. Human population pressures, weather and geology had changed the collections of plants and animals which lived on each island so that they were no longer identical, but represented fragmented parts of the same ecosystem. In the previous three hundred years the peoples of these islands had done their best to restore forests and to recreate extinct life-forms which had vanished during peak human population. However there were still differences and gaps in the way life was represented on each of the thousands of islands.

The first parts of Davana Terra to have life added to them were the northern Beta Mountains and the Themis Regio islands far to the south. The decision was taken that as most of the continent was on or very close to the equator, life should only be added to the coolest part and the terraformers could watch to see how quickly it spread naturally across Davana and to other islands in the archipelago.

Therefore both Themis and Beta were seeded at the same time, initially with figs, cyanobacteria, blue-green algae, lichens, ferns and western Bornean ophiolite volcanic microorganisms. As had been done on Aphrodite, all life forms from the smallest to the largest had to be modified to make them more heat-tolerant. This gave them a better chance of colonising the hot equatorial centre of the Davana continent than would have been the case if their DNA had been left unmodified.

Once again cracks, gaps and gullies were the primary seeding sites for life as they provided shade from the blistering intensity of the mid-day sun. Additionally each site was enriched with a small parcel of nutrient rich fertiliser and water needed to be easily available. On Earth most figs only germinated if they are above the ground so as to avoid competition with other plants. On Venus this was not necessary, as in the absence of plant competition and to avoid the sun's intensity being planted in a crack or gap in the lava lead to better germination success and quicker growth rates than being introduced higher-up.

As more rain arrived on Davana many small rivers appeared and helped to drain the upland areas. Once these had established their courses, life from the source regions was added, including insects, other invertebrates and fish.

About fifteen years after the first figs had been planted fig wasps were introduced, again using heat-tolerant night flying species which had had their trachea enlarged so as to carry away more heat. These were followed by secondary pioneer plants including lupins, desmodiums, myrica and macarang, all good at colonising bare ground and most were also good at enriching the soil with nitrogen and phosphorous. After this more modified microorganisms were added to ash and gravel to continue turning it into soil. Additional soil invertebrates were added one year later.

The next stage of plant seeding was introduced fifteen years later and included thousands of species of plants from across the source region. This happened at the same time as the introduction of the more insects, many of which ate fig leaves and were able to create a food resource for additional bird and bat introductions.

As the years passed it was seen that vegetation was being seeded right across the equator principally by night flying bats. Birds were able to help with this, but the heat killed any which flew south of the Beta Mountains. Nocturnal species could fly across bare rock and lava, but day flying bats and birds were restricted to flying short distances around and between existing patches of forest.

Once the forest groves began to merge around the Beta Mountains, the time was considered right to introduce another set of invertebrate species such as ants, butterflies, bees, moths and spiders. Many of these helped to fertilise forest flowers and so there was more fruit for the birds and bats. Later the first mammals such as monkeys, orangutans, squirrels and slow Loris were added to help with seed dispersal, and larger herbivores like rhinos, deer and elephants were added once Beta's forests were truly interconnected.

Terraforming of the Beta Mountains was relatively slow even though mountain elevation kept temperatures down and made some introductions much easier. Themis however was transformed from bare lava and ash to forest far faster. This was because Themis was further from the equator than Beta and was cooled by both air and ocean currents, which kept the islands below 35-degrees Celsius all year round.

Fifty years after the large herbivores had been introduced; predators were added to Davana to control the numbers of prey species, including snakes, turtles, gharials, leopards, tigers, bears and various jungle cats.

7.3 - EISTLA TERRA

Eistla Terra was a landmass created by the spin correcting meteor impacts in 2680. It was small in comparison with Aphrodite and Devana at just 2,000 kilometres long and 1,200 wide, and lay right on the equator. Therefore it was hot all the time with no peninsula or neighbouring island on which life could get a hold before attempting to migrate to hotter regions.

Initially the Congo rain forest was considered as a possible source of life for Eistla, but it was too hot for even life from Africa's equatorial heart. Day time temperatures in summer were almost 60-degrees Celsius, and only declined to 30 degrees in midwinter.

This meant that no single environment on Earth was a good source for this superheated island and this made the terraformers to begin thinking about terraforming environments in which humans couldn't live.

There were locations like this on Earth, like the super cold Antarctic, and the hearts of baking deserts where people couldn't survive for long. However Eistla was so hot that new sources of heat-tolerant life had to be explored.

There were five possible sources of this sort of life form. The first was based on the few forms of life which had migrated to the equator on Aphrodite and Devana. This limited the introductions to microorganisms, low growing plants, insects and a single species of lizard. However these were hardly sufficient as a resource sufficient to terraform a whole continent.

The second approach was to explore Earth's extreme temperature environments for life which could endure very high temperatures and to use these as a source for DNA which could help the terraformers to build new sorts of organisms.

The third approach was to sift through the Earth's genetic treasure sourced from all the life forms which had had their DNA collected in the previous centuries for heat resistance qualities in parts of their genetic codes. Some sections of genomic data had been inherited from the ancestors of these plants and animals which had lived tens or even hundreds of millions of years before.

The fourth approach was to search Earth's museums for preserved skins, bones, teeth and tusks which had been kept as art or examples of extinct creatures. This did include some heat-tolerant species which had become extinct due to human pressures before genetic sequencing had become possible.

The fifth and final approach to finding useful genetic sequences was to search Earth's fossil record for DNA. This was extremely difficult but was just about possible, as the rare bits of genetic code which were discovered were in such a fragmented state that it needed artificial intelligent machines to piece it all back together again.

The result of this was that the terraformers were able to add more plants to the equatorial regions of Eistla, Aphrodite and Devana, and the implication was that, given time, new sorts of animals would be added as well. However, there remains an inherent danger that these artificial organisms may cause problems for the rest of the life on Venus and so a quarantine area has been developed on a mid ocean Equatorial Island which allows each heat-tolerant life-form to be tested prior to introduction.

8 – OCEANS

Like Mars, Venus's oceans were nutrient rich. On Mars a small number of very large rivers had been the primary contributors to nutrient distribution, but on Venus it was the thousands of small rivers bringing nutrients from the weathering of its mountains which were the key drivers. This high concentration of nutrients had been the case on Mars where the weathering of its southern land mass had brought huge quantities of nutrients into its northern ocean. The implication of this nutrient availability was that, provided life was not buried by constant supplies of sand and silt, life would flourish in the oceans on either planet.

Additionally, the sun's heat and the intense evaporative replenishment of rain clouds meant that Venus was a much wetter planet than Mars with both heat and rain being the daily norm across the whole of the planet. The greater temperatures also drove life to spread around the planet with a speed which had not been the case on Mars.

Consequently, the high temperatures, high rainfall and high nutrient availability encouraged life to grow faster on Venus than it did on Mars or Earth. The increased levels of volcanically produced carbon dioxide also fuelled this process and allowed plants to grow larger and quicker than on either of the cooler planets.

This growth acceleration applied in the oceans as much as it did on land, with fast growth affecting plants such as kelps and seaweeds - as well as corals, which included internal algae to supply the coral with sugars. Provided the water column was clear enough to allow light through, kelp forests and coral reefs could grow to enormous sizes.

When the oceans had been seeded on Mars, plankton rich samples of

Earth's sea waters had been collected, frozen and shipped to Mars. On Venus this was not necessary as seeding and planting shuttles had improved dramatically in the intervening centuries, and many life forms needed to be genetically modified to make them more heat-tolerant prior to introduction anyway.

It was therefore the case that, once an ecosystem had been selected for introduction by the terraforming teams, the autonomous shuttles were able to choose the best place and time for introduction and to make the necessary changes to each organism's genetic structure to ensure its readiness for life on Venus. Besides choosing the ecosystems to be introduced, the only other key role humans played was deciding when each species should be added. For plants this was normally dependant on the availability of water and nutrients, and for animals on the availability of food. However shuttles were beginning to take over this part of the role as they were being upgraded to make species identifications without human guidance.

The other factor which made oceanic seeding easier than land seeding was that Venus's oceans were cooler than the land. This meant that far more ocean introductions were successful with less die off than had happened on the continents.

As had been the practice on Mars, the majority of new oceanic life forms were seeded in their microscopic planktonic form in the ocean's surface layers. However, unlike Mars, the seeding was done one species at a time, as they were printed by the shuttles, thereby making it impractical to seed hundreds of species at the same time. The fringes of the polar continents were seeded first with phytoplankton to begin making a food resource for the zooplankton.

The first sites to be seeded were the oceans around Tethus and Ishtar which were seeded with life from Madagascar, as well as life from parts of the Comoros, Seychelles, Reunion and Mauritius marine ecosystem. This included the gradual introduction of many species of phytoplankton, followed by more advanced zooplankton life forms including corals, jelly fish, sea grass, starfish, sea urchins, crustaceans, sponges and fish. Thereafter reptiles including sea turtles and sea snakes were introduced. Latterly more advanced animal introductions had to be nurtured by robotic mothers including dugongs, dolphins and whales.

The second oceanic seeding was done in the seas around Lada in the southern hemisphere. This was more complex as there were two sources of marine life which could have been used. These were the Pacific on Costa Rica's western coast and the Caribbean on its eastern side. Eventually the decision was taken to use the Caribbean and, more specifically, the Mesoamerican reef which had recovered after being damaged by climate change and had grown southwards to border Costa Rica over the intervening centuries. The decision had been made because the Caribbean was richer in nutrients than the Pacific Ocean and more closely matched the nutrient availability in Lada's seas than the nutrient poor Pacific. Again, it was phytoplankton which was seeded first. This was followed by zooplankton including the larval stages of many species of corals, crustaceans, molluscs, marine worms and starfish. Fish and sea turtles were added later. Other species which had to be nurtured by robotic mothers included manatees, crocodiles, whales, and dolphins.

The third oceanic seeding took place in the waters of the first equatorial continent to have its fringes seeded with oceanic life. This was Aphrodite, which had life from Brazil's eastern coast added to its shores, matching its Amazon rain forest land seeding. The mouths of Aphrodite's larger rivers were seeded with life from the coral reefs directly opposite the Amazon River's mouth. These were not particularly diverse, but the species present were able to cope with extremely rich nutrient supplies and with regular coverings of sand and silt. The primary species present in this assemblage were algae, corals and sponges, but crustaceans and reef fish were also abundant. Thereafter Aphrodite's coasts were seeded with life from three sources; the Abrolhos National Marine Park, the Fernando de Noronha Marine Park and the Manoel Luis Marine Park. Together these represented life from volcanic sea mounts and inshore waters, including corals, sponges, crustaceans, jellyfish, octopuses, shellfish, reef fish, rays and sharks. Later whales, dolphins and turtles were seeded as top level marine predators.

Devana was the fourth marine seeding site, for which life was sourced from the Coral Triangle which included all the land between the Solomon Islands in the east, Sumatra in the west and the Philippines in the north. Many areas within this triangle were included in the collection; however the epicentre of species diversity was the Raja Ampat through-flow off the northwest tip of New Guinea where a large area of small islands and reefs allowed one of the most diverse oceanic areas on Earth to prosper. Species which were specifically included were the argonaut and nautilus members of the octopus family and marine turtles, as well as the larval stages of land crabs which could go onto colonise the forested sections of Devana and Themis Regio. This region matched the rapid oceanic flows which were common around the edges of Devana and its islands, and so Raja Ampat was a good source of life for Devana.

The remaining continent, Eistla Terra, did not have its oceans seeded as the terraformers wanted to know what would arrive on its fringes without any human help. However, there were many other islands, seamounts, shoals and reefs where the waters were well-oxygenated and life would arrive whether it was seeded or not. The three largest of these; Alpha, Metis and Niobe; were seeded on land and in the sea with life from Hawaii, the Galapagos and the Kingman Reefs. This allowed the rare and exceptional forms of life which had evolved on and around these islands to find new homes on which they could continue evolving to suit the conditions they found on Venus.

A number of deep ocean volcanic vents which were similar to the black smokers in the deep oceans on Earth were the last oceanic sites to be seeded in Venus's oceans. This involved the development, programing and manufacturing of deep sea seeding shuttles which could manage the whole process without needing to come back to the surface for months, or even years, at a time.

With the experience of terraforming both Mars and Venus, the terraformers were able to compare the ways in which the two planets had been transformed by the addition of life. In this regard, it was seen that life survived better on Venus than on Mars. Nearly 70% of all species introduced to Mars had become extinct, whereas on Venus the extinction rate was just 6%. This huge difference was believed to be partly due to the more appropriate sourcing of life from volcanic landscapes which had the right microorganisms to help enrich key elements and minerals necessary for life on Venus, partly due to the greater energy input of a hotter, brighter sun, partly due to the greater availability of carbon dioxide and partly because the terraformers worked hard to avoid ecosystem mixing on Venus.

9 - HUMANS

Venus, with its heat, water and earthquakes was a difficult place to live. Keeping anything dry was hard, and almost everywhere on the planet was hot, day and night, winter and summer. Like the natives of jungles on Earth, most colonists took to wearing as little clothing as modesty would allow. This generally involved a short skirt and chest covering; anything more was just too uncomfortable, wet and sticky to endure.

The same conditions which influenced human clothing also affected homes, food and technologies. Thus built structures were quickly damaged by water and heat, and were soon dispensed with once people emigrated from the original colony on Ishtar Terra. A combination of living and non-living composition worked best, with organic structures providing comfort and inorganic elements providing stability and functionality.

Over time homes which had originally been built as linked boxes on stabilising stilts were replaced by mega-dwellings in over-grown trees. These tree-cities allowed the residents to manage heat, humidity and moisture, which would not have been easy to do in a stilt town. The tree itself provided food, water and comfort; and the inorganic structure provided stability, airflow and communication.

The living elements of a tree-city were primarily responsible for the provision of food and water, which meant that fruits were available when needed in special fruiting areas, and water could be found at sip points. The tree also provided its inhabitants with water for washing and cooling, so the plant's transpiration system allowed people to have showers and emitted a soft fog of cooling water vapour in rooms, corridors and on stairs.

The inorganic parts of a tree-city - which were made of nano-robots -

provided the structure which allowed the plant to grow much taller than it would have done naturally. They also built hollow tubes as part of the tree's trunk, through which warm air and water vapour could rise to cool the spaces below. This caused a gentle breeze to waft through rooms and corridors which cooled people by evaporating the moisture on their skins.

The organic elements of the tree also grew resources which humans could use, such as toothbrushes and sanitary paper. The inorganic parts created items like cutlery and cups. All items were recycled by disposal through the tree's waste recycling system centred on the tree's toilets. These fed all of the waste products to the tree's roots where organic components were recycled by fungi which released nutrients and minerals to the tree's roots, and inorganic components were reused by nano-robotic reabsorption into the tree's skeleton. These tree cities were also able to absorb the carbon dioxide their citizens breathed out, giving the plant additional resources to enable them to grow taller and stronger.

Lighting in these trees was provided by fluorescent bacteria which were painted onto walls, steps and floors. Here they grew in the warm wet conditions and allowed humans to become partially nocturnal; rising early, enjoying long siestas during the heat of the day, and staying up late into the night. As this pattern was a cooler and more comfortable way of living, it became increasingly popular as the decades passed.

Some things were very difficult on Venus; such as keeping anything made of paper or wood in good condition. So, paper books and antique furniture shipped from Earth rapidly deteriorated and had to be stored in special conditions. Mostly people were happy to see these things virtually – in their mind's eye – rather than having to maintain the real thing.

By the mid-2900s, human minds had been supplemented by the addition of organic computer elements which gave people superior intellect and advanced analytical capabilities, but also meant that they needed to be able to share their thoughts with others without interruption. The addition of organic radio frequency detectors and generators to people's brains meant that people could communicate through thought alone. These thought collectives allowed people to work on projects remotely with thousands of others, without breaking from their normal routines. Working whilst lying still was a very popular pastime; meaning that no one could tell if a person was asleep, just resting or actively trying to solve a communal problem.

Additionally, as the supplementing of people's minds with living computer components continued to progress, the technology allowed humans to control machinery and processes without the need to see or be close to it. This meant having a general awareness that all was well was sufficient for colonists, with people only needing to pay attention when something went wrong or the system needed an upgrade. Because of this - and due to the help provided by robotics, offices, laboratories and workplaces gradually disappeared as people became more familiar with mixing leisure, discussion, research and work on a minute-by-minute basis.

Being wet much of the time did cause people health problems such as fungal infections and allergies. Being able to 'air' the skin allowed both heat and moisture to dissipate. Additional help was provided by "living clothes" which had the ability to channel moisture away from people's bodies. These clothes had to be worn in the light and immersed in water each night to keep them alive. The walls, floors and surfaces of people's homes and offices were also able to suck moisture away, but used tiny hairs which moved air so as to remove moisture from people and their living environment.

Just as the stilted towns had disappeared after people moved away from the founding colony, stilted factories were also superseded by tree buildings. As most manufacturing was done using nano-robotic machines, and anywhere humans needed to work had to involve evaporative tree cooling, tree factories were grown after the year 3000 to allow people to enter and monitor antonymous manufacturing facilities.

Other things which were grown on Venus included sun screens and shades, hats, clothes, shoes, and of course food. In some places sun shades have been grown to cover complete towns, transport plate stops and walkways. Lone bio-nano trees were grown as living communication hubs which gave increased connectivity so that people didn't lose touch with the mind net.

This combination of biology and micro-robotics also found uses in the rest of the solar system where plant-nano partnerships - developed on Venus - could be used for keeping the air fresh, providing food, pumping water and communication.

On Venus the best way to make short journeys was by anti-gravity plates which were two meter wide disks stacked ready for passengers at plate stops. They used nanos to quickly grow seats and safety barriers to prevent people from falling off, and rain covers could also be quickly grown when the weather required it. The plates included communication hubs so that passengers didn't lose contact with the rest of the mind net on route. For longer journeys people caught shuttles which could take the traveller between cities or across oceans.

With human nature being what it had always been, some people ate too much or failed to exercise. However, membership of body feedback organisations helped members to achieve complete control of their bodies and gave them better employability and health outcomes. People who gained or lost too much weight were regarded as unwell and sent for treatment in clinics where their motivation and determination to attain a healthy weight were restored before they could gain access to many societal benefits such as luxury foods and holidays.

For relaxation people played sports, and some took to sailing and hiking. With a rapidly growing young population, the agenda was set by people who wanted to be fit and to have an active lifestyle.

Continuing earthquakes and volcanism limited where people could live and work, but many were keen to test the bounds by growing quake-proof tree dwellings and having escape plates close at hand. Nevertheless Venus could be unpredictable and autonomous emergency plate rescue was a major feature of living on a volcanic planet.

With nano-robotics doing almost all of the difficult or boring jobs, people had much more spare time and found that the dividing line between work and relaxation was hard to define. Most people had jobs which involved designing nano-robotic machines or modifying DNA to create living things, and sometimes both. Then the dividing line between their working with the collective mind to create something new and recreation were indistinct. So it was that work and play became part of something greater which was more akin to an all-consuming obsession than it was a traditional job. So people hiked for recreation across mountains and through forests, and at the same time designed new products and researched the potentialities of sections of DNA.

As had been the case for hundreds of years, people still found pleasure in eating, and restaurants became as common on Venus as they were on Earth. The same applied to fashion and clothing; so that retailers and shopping areas became common. The regulations stated that any material or substance could be sold provided that it could be recycled by people's home recycling systems.

Whilst there were mining, gas collection and farming operations on Venus, these processes were almost entirely automatous, being run by nano-robotics and managed by intelligent machines. Humans were only involved in regulating these processes to prevent damage to the environment and harm to humans. These operations were, however, how Venus earned most of its foreign currency so that goods could be imported and Venusians could afford to visit Mars or Earth.

With improved shuttle engines and shorter transit times, both Mars and Earth were just weeks away so that, whilst they were major journeys, the whole holiday or research trip normally only took a few months. Additionally, with its similar gravity, visiting Earth was much more comfortable for Venusians than it was for Martians who suffered even in their motorised support suits. This meant that visitors from Venus could hike on Earth's mountains, and tourists from Earth could explore Venus's forests.

People living near the equator endured rather than enjoyed the warm conditions and did everything they could to reduce the negative effects it could have on their children. Some parents began giving permission for clinics to modify the genetic structure of their unborn children. Ethicists did not at all like this trend but, with the number of people dying of heat exhaustion on the rise, the trend toward human modification continued.

Children on Venus lived with their parents. This was different from Earth where children for the most part lived in birthing centres. The cause of the move back to more traditional ways of bringing up children was the need to populate Venus and people thought of themselves as pioneers continuing the terraforming process. This dual focus on the future of Venus and on historical family structures and values struck many people on Earth as strange. However on Mars and Venus it made perfect sense.

Protecting people from the dangers of genetic damage due to solar and cosmic radiation continued to be a problem. The solar magnetic shield which was placed between Venus and the Sun in 2933 provided protection against ionised particles coming from the Sun, but this did little to protect people against the damage inflicted by radiation coming from the rest of the cosmos.

A number of strategies were started to try and achieve protection from cosmic radiation, the first of which was the placement of additional magnetic shielding generators around the rest of the planet. Thereafter the ozone layer was supplemented using floating solar-powered micro-factories which made ozone from normal oxygen, ultraviolet light and an electrical spark. These devices also helped reduce surface temperatures around the equator.

Since the first people had landed on Venus's surface there have been political and legal problems with land ownership. Land rights had initially been apportioned by the authorities on Earth, but people took little notice of these due to the amount of land available for the small Venusian population. There were difficulties with the founding of tree-cities, as some had been started in areas other citizens wanted to be nature reserves. This led naturally to the introduction of land use registration which ensured that, once registered as a city, wild area or mineral mine, no changes could take place without court approval.

Other legal changes which distinguished Venus from Earth and Mars were the outlawing of manufacturing or owning of weapons or ammunition, which made disputes about land and property more peaceful than could otherwise have been the case. These laws also ensured that nature was allowed to build its own predator-prey relationships without hunters upsetting the balance.

Although many species were considered sustainable and could have been hunted for their meat, most people on Venus were vegetarian simply because the advances which had been made in genetics and growing food meant that protein was available in tasty-to-eat plant-form almost everywhere. With fruiting rooms in tree cities, at plate stops and in shopping areas, if people wanted some meat-like food they could either pluck a berry, or they could go to a restaurant where more unusual plants and spices were served.

On the coasts of Venus's continents fish was available, which was a great novelty for most citizens. However, as it tasted far better fresh than it would have done had the colonists risked the use of refrigeration in Venus's high-temperatures; it was regarded as a holiday treat by most citizens. Some of this fish was caught by fish-bots which could identify the right species; age and condition of a fish the rest were caught be people who liked to catch fish the old way with rod and line.

Legacy livers were reasonably common on Venus, much as they had been on Mars. People who wanted to live on their own - away from the rest of the population - doing things in ways they had been done hundreds of years before were considered rather exotic and strange. They needed special legal permissions to live in some areas, but were generally accepted as being harmless eccentrics.

People on Venus had more leisure time than the populations of the other planets, and could follow their interests rather than having to work to live. This meant that Venus's population was highly creative and produced large numbers of innovations, particularly in combining nano-robotics with living things.

The gradual change from traditional employment to exploration, creation and discovery had been forced on the populations of all three planets by advances in robotics and artificial intelligence. These changes meant that people were free to follow their own interests rather than having to spend their lives working in a job they hated.

As children, people were paired with virtual intelligent robots which helped them progress through the educational system. These robots had no set curriculum and would instead allow a child to ask questions all day and all night, resulting in generations of children with interests and enthusiasms in subject areas that human educators might not have considered or allowed.

This process continued into adulthood, although most people gradually changed their relationship with the virtual robot until it became just a way of accessing information. Nevertheless, as the robots had the ability to analyse and suggest, they were still important aids - an artist might paint a picture and ask his robot to see similar images to inspire his imagination, an engineer might design a machine and ask to see similar designs, and a geneticist might develop a stretch of DNA and ask to see similar genes with the same function.

This ease of access to relevant and correct information meant that people could learn very rapidly from the experience of others, with the virtual assistant acting as if it were part of the person's own brain. It also meant that truly creative minds were able to earn more credits than people who were less original.

Everyone on Venus had access to a basic level of resources to keep them alive. No one went hungry, or thirsty, or was without a place to live or clothes to wear. Products classed as basic needs were freely available, although a record was kept of overconsumption. Only products defined as luxuries were restricted, and these were available in exchange for creativity and originality.

This drive for originality and the use of virtual robotic teachers meant that people all had different knowledge, interests and abilities. There was no such thing as an averagely-educated person or a standard educational curriculum. It also meant that people with similar interests could be a long way apart geographically and needed to be able to share thoughts and knowledge in a quick and easy manner.

In some areas tree-cities were populated only by people with the same interests, so temporary towns near the equator were populated by geneticists with an interest in life at high temperatures. During the summer they migrated back to their town of origin, but other interest communities were more permanent.

So individual towns around the planet specialised in cryptography and others in nano-robotic animals; in fact living with people with the same interests was popular on each of the three planets. About thirty percent of Venus's towns had a specialist focus; and this proportion grew till it reached about fifty percent.

10 – MACHINES

Venusians were competing with Mars and their development of nano-robotic ecosystems in which machines could design, build and repair other machines. These were used extensively in mining the asteroid belt and the moons of Saturn and Jupiter. Venus's inventiveness – stemming from its carbon dioxide availability – was mainly in making living organisms do useful things for humans such as growing permeable covers over towns or making interplanetary ships more comfortable with living walls, fruiting areas and waste recycling.

The first dronic ecosystem was developed on Mars so that machines could design and build other machines. This made mining and manufacturing far easier on each of the three planets. However it also brought with it fear that these machines would go feral and begin competing with humans. Therefore dronic ecosystems were only allowed in areas without human habitation. This included many of the moons in the solar system, the asteroid belt, and ultra-deep mines on Mars and Venus.

Together these dronic machine ecosystems were seen as a precursor to the arrival of people, but in practice many were set to work in places humans would never go. They were able to assemble places for humans to live in inhospitable circumstances well before humans arrived to begin work. They were, for example, used to create colonies on Enceladus and Europa, which meant that the construction was far faster than expected and no one died in the process.

Another use of these machines was the monitoring of life on Venus's equator, where it was too hot for people to survive during the summer. Here they were able to monitor the introduction of species as diverse as bats and lizards to understand their movement and survival during times of high temperature and humidity. As these machines were able to adapt to changing conditions they were able to follow the bats into caves and the lizards into the ocean.

The deeper parts of the ocean were another place on Venus where self-replicating machines were able to do things people could not. Biologists wanted to know which species were surviving at depth, what they were eating and whether newly introduced species were able to compete in individual niches. The great advantage that self-replicating and self-designing machines had was that they were able to find ways to survive, despite facing adverse conditions which humans might not have anticipated.

The same methodology will in the future apply to terraforming newly discovered planets in other solar systems, such as bringing the necessary liquids and gasses to the new planet, altering its day length, tilt and even year length before humans needed to even approach the planet. They might also add moons to a newly terraformed planet and introduce life to make it hospitable before humans arrived.

In the future this is probably how terraforming will happen, with a dronic ecosystem self-replicating and designing new machines to prepare a planet for human arrival. All that humans would need to do would be to select which of Earth's ecosystems could be used as a template for a new planet and set the rules which would control the sort of habitation to be built for them.

11 - CONCLUSION

Questions remain about Venus. For example, could we have changed Venus's year length and could we have added a moon? These are questions which may be looked at in the future if new sorts of technologies become available or if new planets are selected for terraforming.

There is no doubt that a moon would make Venus a more biodiverse planet by adding wide inter-tidal zones to most sea shores. However, there is always risk in this kind of mega-scale project.

Another project which has been contemplated is the creation of world ships sharing an orbit in the asteroid belt. These would be created using resources sourced from the asteroids and would take the form of huge cylinders, spinning so as to create artificial gravity. This would allow people, plants and animals to live on the internal walls of each cylinder. This region of space, between the orbits of Mars and Jupiter is far colder than humans could endure. In spite of this the plans suggest that fusion air warmers could be used to create comfortable conditions in each world ship.

As humans we are strange beings, in turn builders, gardeners and destroyers. When we first meet an alien intelligence, what will they think of us?

Milton Keynes UK
Ingram Content Group UK Ltd.
UKHW021516170524
442874UK00033B/257

9 780995 674158